# FUTURE ENERGY

Gas goes across the earth in great pipes. Oil and coal travel from one country to another in big ships, often for thousands of miles. We do not think about this when we take a cold drink from the fridge, or turn on a light. Energy has always been there when we wanted it.

But the clock is ticking. The oil, coal, and gas will not last forever. Scientists are working hard to find new ways to get energy, and some of their ideas will surprise you. A car that sails with the wind, a turbine at the bottom of a river, machines that use the heat from people's bodies – these are some of the places that the energy of the future will come from. And that future is not far away . . .

OXFORD BOOKWORMS LIBRARY
*Factfiles*

# Future Energy
Stage 3 (1000 headwords)

Factfiles Series Editor: Christine Lindop

ALEX RAYNHAM

# Future Energy

OXFORD UNIVERSITY PRESS

# OXFORD
UNIVERSITY PRESS

Great Clarendon Street, Oxford, OX2 6DP, United Kingdom

Oxford University Press is a department of the University of Oxford.
It furthers the University's objective of excellence in research, scholarship,
and education by publishing worldwide. Oxford is a registered trade
mark of Oxford University Press in the UK and in certain other countries

ISBN: 978 0 19 479449 7

A complete recording of *Future Energy* is available on CD. Pack ISBN: 978 0 19 479448 0

Printed in China

Word count (main text): 10,244

For more information on the Oxford Bookworms Library,
visit www.oup.com/elt/gradedreaders/

ACKNOWLEDGEMENTS

*Cover image*: Corbis (solar panels/Tetra Images)

*The publishers would like to thank the following for their permission to reproduce photographs*: Alamy
Images pp.4 (Coal miners/Miguel Sayago), 24 (Petrol pumps/Sandy Young), 26 (Solar cooking/
Joerg Boethling), 33 (Itaipu hydroelectric dam/Mike Goldwater), 37 (Turbine/Frances
Roberts); Andrew Engineering Ltd p.39 (Ground loop); Corbis pp.7 (Fuel tanker ships/Tim
Wright), 11 (Rusting oil barrels/Ashley Cooper), 12 (Heavy traffic/David Butow/CORBIS
SABA), 16 (Nuclear storage pond/Ocean), 22 (Sugar cane harvest/Tim Page), 27 (Solar-powered
airplane/Denis Balibouse/Pool/epa), 28 (The PS20 solar tower/Ashley Cooper), 31 (Bahrain
World Trade Centre/Hamad I Mohammed/X01444/Reuters), 41 (Human-powered Daedalus
Aircraft/Charles O'Rear), 50 (1973 Opec Oil Crisis/H.Armstrong Roberts/ClassicStock),
54 (Crowds in Shanghai/Imaginechina); Maybach pp.52 (Maybach DRS), 52 (Maybach DRS);
Oxford University Press p.32 (Windmill); Peter Lyons p.32 (Greenbird wind powered car);
PlayPump p.42 (PlayPump roundabout); Rex Features pp.14 (Freegans/John Alex Maguire),
18 (Damaged Fukushima Dai-ichi Nuclear Power Station), 29 (Wind farm/Invicta Kent
Media); Science Photo Library pp.0 (Seoul, South Korea, at night/NASA), 2 (Steam engine/
Jean-Loup Charmet), 5 (Oil rig/F. Ayer), 8 (Deepwater Horizon oil rig fire/USCG ), 9 (Coal-
fired station/Martin Bond), 15 (Landfill site/Martin Bond), 20 (Fusion reactor maintenance/
Maximilian Stock Ltd), 23 (Microalgae production/Matteis/Look at Sciences), 35 (Pelamis
wave power/Ocean Power Delivery/Look At Sciences), 38 (Blue lagoon geothermal pool/
Martyn F. Chillmaid), 44 (Medical nanorobots/Hybrid Medical Animation), 46 (International
Space Station construction/NASA), 52 (RepRap machine/James King-Holmes).

# CONTENTS

Seoul, South Korea, at night

# 1        Energy today

I can feel the plane shaking as it gets faster, then suddenly lifts into the air. Its engines are burning 1.5 litres of fuel every second as we climb into the sky above Istanbul. It is seven o'clock on a cold February evening in a city of 13 million people. Below me, people are travelling home from work in cars, buses, trains, and boats.

Through the plane window, I can see thousands of lights from factories, streets, shops, houses, and ships in the Marmara Sea. The city lights look beautiful at night, but have you ever thought about how much energy they use? Where does all this energy come from?

About two hours later, I open the front door of my house in Adana in eastern Turkey. I can smell food cooking, and hear the sound of a TV. A red light goes on and off on the telephone. All over the house, machines are taking messages, washing, cooking, and heating the house.

Perhaps you are reading this book at home. Are the lights on in your room? Are you listening to music? How many machines can you see around you right now?

At this moment around the world, billions of lights, computers, TVs, and fridges are turned on. At any moment of any day, 25 million cars are driving on roads and nearly 700,000 people are flying somewhere in a plane. Most of the energy that we use for these things comes from fuels like coal, oil, and natural gas. One day soon, we will not have any more of these fuels. Where will our energy come from in the future, and how will this change our world?

# 2    Fossil fuels

For thousands of years, people made things with their hands. They used the power of the wind, water, and animals to travel, move, or build things. Most people burnt wood to heat their homes and to cook. Then, in October 1765, a young engineer called James Watt built a machine that changed the world – a steam engine.

To make a steam engine work, coal is burnt to heat water, and this makes steam. The steam goes into the engine and moves the parts inside it. There were steam engines before 1765, but Watt's new engine worked much better and could move big machines in factories. Soon people began to build factories everywhere.

Watt's steam engine

In the next hundred years, lots of new factory machines were invented. They made new products for people to buy. Suddenly our houses were full of new things. In many countries, thousands of people left their villages and moved to the cities to work in the factories. Poor workers worked long hours with dangerous machines, and life was very hard for them. Smoke from the burning coal filled the air in the towns.

The first steam train was built in 1804. By 1850, trains and ships with steam engines were carrying passengers around the world. In the 1880s, the first power stations were built. They burnt coal to make steam for huge machines called steam turbines. When steam turbines move, they turn parts in machines called generators, which use this movement to make electricity. Soon electric lights appeared on the streets and people had electric power at home.

In 1885, German engineer Karl Benz invented the first car. It used a new type of engine and a new type of fuel: petrol. Petrol burnt inside the engine to make the parts move, and this made it much smaller than a steam engine. Twenty years later, factories were producing thousands of cars every year, and new roads crossed the land. Most of these cars used petrol, which comes from oil. In 1903, Orville and Wilbur Wright used a petrol engine to fly the world's first aeroplane.

Things like the steam engine, electricity, and the car changed the lives of everyone on earth. They also changed the way that we get energy. Today about 87 per cent of the world's energy comes from burning coal, oil, and natural gas. Where do these fuels come from, and how do we use them today?

Coal, oil, and natural gas come from things that were alive millions of years ago. Oil and natural gas come from animals that lived in the sea. Coal comes from plants that lived in wet places, such as land next to rivers or lakes. Over millions of years, far under the ground, they changed into coal, oil, and natural gas. We call these kinds of fuel fossil fuels, and the oldest ones are about 400 million years old!

## Coal

We have used coal for a long time. Two thousand years ago, the Roman people used it to heat their homes and make metals. We still use coal for these things today, but most coal is burnt in power stations to make electricity. About 40 per cent of the world's electricity comes from coal. Every week, somewhere in the world, a new power station that burns coal is built!

To make the electricity that your fridge uses in one year, you need about 300 kilograms of coal! The biggest coal power stations burn 10–15 million tonnes of coal every year. A lot of that coal comes in ships from thousands of kilometres away.

In some places, we get this coal from huge holes on the surface of the earth. In other places, the coal comes from hundreds of metres under the ground. There is not much room to move, and the temperature can be 40 °C or more.

Coal miners at work in Chile

It is difficult to get enough clean air and often too noisy to speak. Getting the coal from under the ground is dirty, dangerous work, but millions of people do it every day. Every year, about 5,000 of them die.

## Oil

In places like Saudi Arabia, Nigeria, and Venezuela, there are lakes of oil, called oil fields, under the ground. To get the oil, people drill holes in the ground called oil wells. Some of these wells are several kilometres deep! In other places, huge machines called oil platforms drill wells under the sea. There are often bad storms at sea, so oil platforms have to be very strong. Under the water, some platforms are as tall as skyscrapers – the world's tallest buildings.

An oil platform
in the North Sea

Countries with oil fields send the oil to other countries in long pipes, or in huge ships called oil tankers. The world's biggest oil tankers can carry 440 million litres of oil – that is as heavy as 350,000 family cars!

Oil contains many chemicals. The tankers take it to factories where it is heated and cooled to get the different chemicals from it. Some of these chemicals are used to make things like plastic or clothes. However, about 85 per cent of the oil is made into fuels. There are different types of fuel for engines in cars, ships, and planes. The factories also make fuel for heating buildings, and for burning in power stations to make electricity.

Every year, we make about 60 million new cars, and thousands of ships and planes – so every year we need more and more oil.

## Natural gas

About 2,000 years ago, people in China made pipes from tall bamboo plants. They used them to drill wells and find natural gas hundreds of metres under the ground. The pipes carried the gas to their homes, where they used it for gas lights and heating water.

Today we burn natural gas in factories and power stations. We also use it in homes for heating and cooking. There are even cars and buses which drive on natural gas instead of petrol. Natural gas is the cleanest fossil fuel: it produces much less pollution than burning coal or oil.

When a person in Britain cooks something, the natural gas that they use may come from Norway, Russia, or Kazakhstan. How do they get the gas from these places? Often the gas goes through pipes. One gas pipe under the sea from Norway to Britain is 1,200 kilometres long! In

Gas tankers in
Virginia, USA

other places, the gas is cooled to make it into a liquid. This liquid gas is put in ships called gas tankers.

A lot of the world's natural gas is found inside a type of rock called shale. In the past, it was too difficult and expensive to get the gas from inside the rock. Now in places like Pennsylvania, in the USA, people are using water to break the shale rocks under the ground and get the gas. In the first ten weeks of 2011, three hundred new gas wells were drilled in the USA. The problem is that each well will produce millions of litres of polluted water. You have to clean all this water or keep it somewhere safe.

It is always the same story. Today shale gas is the newest fossil fuel to bring good things for some people – new jobs and money – and bad things for others – pollution from dirty water. In rich countries, fossil fuels have made it possible for most people to live a very comfortable life. Will they destroy that life one day too?

# 3  Energy and our planet

In the waters of the Gulf of Mexico, between Mexico and the USA, there are more than 2,300 oil platforms. On 20 April 2010, oil workers were drilling on one platform when gas from the well exploded. The oil platform was destroyed and eleven workers died. After the accident, oil started to escape from the well at the bottom of the sea.

It took two months for people to close the well. In that time, nearly 800 million litres of oil went into the sea. Ugly, black oil polluted beaches for hundreds of kilometres. Many birds and sea animals died, and people who worked in tourist and fishing businesses lost their jobs.

Fire on the oil platform in the Gulf of Mexico, April 2010

Gases and steam from burning coal in the UK

As the world's population grows, we need more and more energy. To find enough coal, oil, and natural gas, people are digging and drilling deeper. Energy companies are searching for fossil fuels in places like Alaska and the Amazon. But pollution and accidents can cause great damage to these beautiful natural places.

Burning fossil fuels produces dangerous gases. Some of them pollute our cities and damage people's health. Every year, about 2 million people die because of air pollution. Scientists think that other gases, like carbon dioxide ($CO_2$), are changing the world's climate. If a plane flies from Singapore to Los Angeles and back, its engines produce about 7 tonnes of $CO_2$ for every passenger on the plane.

Since the steam engine was invented, the amount of $CO_2$ in the air has grown by 35 per cent. $CO_2$ catches heat from the sun, so this makes the climate warmer. Because of this, in some places there is less rain than there used to be. Farmers cannot grow enough food, and forests are burning in the hot, dry weather. In other places, there is now too much rain: terrible floods destroy farms and houses.

On high mountains and in the Arctic and Antarctic, warmer weather is heating the ice and snow and changing it to water. This means that more water is going into the sea, so the sea is getting higher. Islands around the world are starting to disappear under the sea. Cities on the coast, like Shanghai, Dubai, and Venice, may disappear one day too.

Many living things are dying because a hotter climate is changing the places where they live. From the forests of Costa Rica to the ice of the Arctic Ocean, the land is changing and animals are disappearing. The facts are frightening: every day the world loses about 150 different types of plants or animals.

At the moment, the richest countries in the world use most of the energy and produce most of the pollution. For example, the USA has only 5 per cent of the world's population, but in any year it uses about 25 per cent of all the world's energy. It also produces about 45 per cent of the world's $CO_2$. Australia produces more carbon pollution per person than any other country. But as other countries get richer, their populations want more things like TVs, computers, and cars – and that means they are starting to use more and more energy to produce and run them.

Around the world we use about 12 billion litres of oil, 19.8 billion kilograms of coal, and 10 billion cubic metres ($m^3$) of natural gas every day. But scientists think that forty or fifty years from now, there will be no more oil. About twenty years after that, we will have no natural gas. Finally, in about 120 years, we will finish all of the world's coal. One day, all the fossil fuels will be gone.

Pollution in Greenland

We do not need to use fossil fuels: there are lots of other ways to produce energy. The problem is that most of the world's car engines, heating machines, and power stations were built to use fossil fuels. Changing this will take a long time, so a lot of people want to try to save energy too. The good news is that there are lots of ways to do this.

# 4                 Saving energy

About half of the energy that we produce is wasted. Electricity is lost in power cables and cars waste fuel as they wait in traffic. Governments can save energy by building better power stations, for example, but we can help a lot too. What can we do to save energy?

## Cars

On the Santa Monica Freeway, in Los Angeles, thousands of people are trying to get to work, but they are going nowhere.

**Going nowhere in Los Angeles**

A driver in Los Angeles spends about 70 hours a year in traffic that is not moving. While long lines of cars wait under the hot Californian sun, their engines are producing dangerous gases. Drivers look angrily at their watches. Sometimes, you cannot see the sun in Los Angeles because of all the pollution in the air! It is the same every morning in São Paulo, Moscow, Bangkok, and many other cities.

What about other ways to travel? Walking and cycling to work or school are great ways to save energy and stay healthy as well. If you cannot cycle or walk somewhere, you can still save energy if you take a train or bus instead of driving. In a lot of cities, people are joining car-sharing groups; each person in the group drives their friends to work one day a week. Everyone saves petrol and money, and there are fewer cars on the road.

## Food

All the time, millions of tonnes of food are moving around the world, and this uses huge amounts of energy. Of course, you cannot grow tea in Iceland or rice in Qatar, so we have to buy food from other countries. It is great to eat Italian spaghetti or enjoy a hot cup of Kenyan coffee, but it is also good to think about where some of the food on our table comes from. If we want to save energy, we can try to buy more things that were produced in our country. In Turkish supermarkets, for example, you can buy bananas from Anamur, in Turkey, or from Ecuador, which is 12,000 kilometres away.

People say 'You should not go into a supermarket when you feel hungry', and it is probably true. Supermarkets are full of fantastic things, and it is easy for hungry shoppers to buy more food than they really need. In European countries

Good food is thrown away every day

each family throws away more than 1,000 dollars of food every year – 20 to 25 per cent of all the food that they buy. Some of that unwanted food has come in ships from the other side of the world. So next time you go shopping, it may be a good idea to eat something first!

## Energy at home

Imagine two power stations working 24 hours a day, 365 days a year. That is how much energy people in Britain waste by leaving things like TVs on standby (turned on and ready to use). Did you know that leaving a TV on standby all day can use the same amount of power as watching it for an hour? How many machines are on standby in your house? Can you turn any of them off?

When you turn on a normal light bulb, only 10 per cent of the electricity it uses turns into light – the other 90 per cent is wasted as heat. Using energy-saving light bulbs saves 80 per cent of that electricity, and you can use them for much longer too.

Do you like to heat your house to 26 °C in the winter or cool it to 18 °C in the summer? Half of the energy that we use at home is used to heat or cool the house. You can save a lot of energy if you keep your house at 22 °C all year and put on warmer or cooler clothes. It also takes a lot of energy to heat water. When you make a hot drink, it is good to heat just as much water as you need, not more.

## Rubbish

Every day, people in Greece throw away 8 million plastic bottles. Around the world we produce millions of tonnes of rubbish every year. Some of this rubbish is recycled, but most of it is thrown into big holes in the ground. Rubbish like this may be dangerous for animals and people for hundreds of years.

Recycling an old aluminium drinks container only uses 5 per cent of the energy that we need to produce a new one. Rubbish is taken to places called recycling centres, where it is put into different groups. Later, each different kind of rubbish is broken into pieces and made into new materials. Recycling is easy to do, it is good for the natural world and it saves a lot of energy. Is there a recycling centre near you?

Saving energy means thinking more about the things that we do every day. If we do this, we can stop a lot of pollution, and save money too!

Rubbish goes into the ground in the UK

# 5

# The power of the atom

On a cold afternoon in December 1951, a small group of scientists stood in a room in Idaho, USA. They watched excitedly as four ordinary light bulbs were turned on, then they shouted and shook hands. They had just invented a new way to make electricity: nuclear power.

Inside a nuclear power station

Today, nuclear power produces about 13 per cent of the world's electricity. One kilogram of nuclear fuel can have as much energy as 1.5 million kilograms of coal! When nuclear power was invented, some people thought it was the answer to all our energy problems. Today, many people are afraid of it. So what is nuclear power, and what are the dangers?

Everything around us is made of atoms. Some metals like uranium are radioactive, which means that the centre of the atom can break. The process of breaking the centre of the atom is called nuclear fission, and it is this process that produces nuclear energy.

Uranium 235, the type of uranium that we usually use for fuel, is found in rocks around the world. It is difficult and expensive to get it from the rocks and make it into fuel. In nuclear power stations, sticks of uranium 235 fuel are put inside a place called a nuclear reactor. Other sticks called control rods go between the sticks of fuel. They stop the reactor from becoming too hot.

In most reactors, water is used to cool the fuel and the water then becomes hot. This hot water moves through pipes and heats 'clean' water outside the reactor. The clean water turns into steam that moves steam turbines.

Energy from nuclear fission travels through other things. This moving energy is called radiation, and it makes everything inside the reactor building radioactive. Radiation is very dangerous for people, so nuclear reactors have thick, strong walls. These stop water, gas or anything inside the reactor from escaping.

About thirty countries have nuclear power stations, and others want to build them. Many people think that using nuclear power is better than burning fossil fuels because it does not produce gases like $CO_2$. Other people worry about nuclear waste and accidents.

## Nuclear safety

Old fuel and other waste from nuclear reactors is very radioactive – and very dangerous. Some waste is recycled and used in reactors again, but a lot is kept in very strong containers under the ground. This worries many people. Could nuclear waste get into water in the ground one day? Could people steal nuclear waste and use it to make a bomb? Some nuclear waste will be dangerous for 20,000 years: that is a long time to keep something safe!

On 25 May 1986, nuclear fuel in a reactor at the Chernobyl power station in the Ukraine began to get hotter and hotter. Workers could not cool the reactor, and finally it exploded. Radioactive fuel and control rods were thrown high into the

Fukushima nuclear power station after the tsunami

air. Fires in the reactor burnt for fifteen days, and radioactive smoke travelled across Europe and the Black Sea. By 2005 fifty-six people had died because of the accident. Scientists think that, in time, about 4,000 people in the Ukraine, Belarus and other countries may die because of Chernobyl.

Today, most nuclear power stations are much safer than they were at the time of Chernobyl, but things can still go wrong. On 11 March 2011, a huge earthquake happened under the sea near Japan. About forty minutes later, a wave 14 metres high hit the coast near the forty-year-old Fukushima nuclear power station. Water destroyed machines at the power station so people could not cool the reactors. In the days after the accident, teams of brave engineers worked day and night to keep the power station safe. There were big explosions, and radioactive gas went into the air. Later, radioactive water went into the sea. Much less radiation escaped from Fukushima than from Chernobyl, but it showed the world that safety in nuclear reactors is still terribly important.

## Nuclear fusion

Around the world, scientists are trying to build a new type of nuclear reactor. If they succeed, nuclear power will be much safer, and it will also become much more important. The new reactors will get energy from nuclear fusion.

Nuclear fusion happens on the sun, and it produces a lot more energy than nuclear fission. Inside the sun, atoms of hydrogen join together to make bigger helium atoms, and this produces a huge amount of energy. Future nuclear fusion reactors will not produce much waste, and they will also be safer because you can stop the fusion process quickly. So why are we not using them today?

Inside a
fusion reactor

Imagine trying to put the sun inside a room, and you will understand how difficult it is to build a fusion reactor. To start nuclear fusion, the atoms in the fuel must reach temperatures of about 150 million °C!

At the moment, people are trying different ways to start nuclear fusion and keep the hot fuel from touching the sides of the reactor, but it is very difficult. In the future, we may power our cities with nuclear fusion, but we have to find answers to a lot of problems first!

While nuclear fuels continue to provide power in many countries, scientists keep looking for new fuels that are clean and safe – and some of their ideas are quite unusual.

# 6       Super fuels

Imagine growing fuel on trees or getting it from rubbish. Imagine cars that run on air or produce water, not pollution! All around the world, people are making surprising new fuels.

## Biogas

Biogas is made from plants or natural waste by living things called bacteria. They break the waste down and produce gases like methane and carbon monoxide. In most places, the gases pollute the air, but at Bandeirantes near São Paulo in Brazil, pipes take the gas from under the ground. The gas is burnt at a power station to produce electricity for 400,000 people!

Villagers in India use animal and food waste to make biogas. Bacteria break the waste down in a special container, and the gas is used for cooking and lights. Just 1 kilogram of waste produces enough biogas to make a light work for four hours. In Sweden, they even make biogas for cars from toilet waste! A year's waste from seventy toilets makes enough fuel for a small car to travel 16,000 kilometres.

## Biofuels

Biofuels are made from fuels that grow. The oldest biofuel is wood, but today we are using different kinds of plants – and even old coffee! – to make new biofuels.

The problem with some biofuels is that it takes a lot of land to grow the fuel, which means less land for growing food. If you want to fly a Boeing 747 plane the 350 kilometres from London to Amsterdam on coconut biofuel, you will need to grow about 3 million coconuts! There is another problem too – people may destroy forests to grow plants for biofuel.

In Brazil, cars and buses have used a fuel called bioethanol for years. Bioethanol is great because we make it from the waste parts of plants that we already grow for food. Most Brazilian bioethanol is made from sugar cane, the tall plant that we grow for sugar. The sugar cane is broken up in machines, and the liquid is taken to make sugar. The rest of the plant is turned into paper or used to make bioethanol.

Cutting sugar cane

Growing algae for biofuel in France

Even the electricity that the bioethanol factories use comes from burning sugar cane. Burning biofuels produces gases like $CO_2$, but when we grow them, the plants take $CO_2$ out of the air. Because of this, sugar cane bioethanol produces 78 per cent less $CO_2$ than petrol.

The newest biofuels are made from very small living things called algae. Algae produce more energy than other biofuels, and they can grow in places where we do not grow food, like seas and waste land.

In the future, people want to change the way that some bacteria and plants grow. In this way they hope to get biofuels that work better or grow in different places. One day there may be huge algae farms in the sea, and many buildings may grow biofuel plants on their roofs.

## Driving on air!

Electric cars do not burn fuel: devices called batteries keep electricity and use it to power the engine. People have driven them for years, and they are a great way to have less pollution in city centres, but where does the electricity come from? It may come from burning coal, for example.

**BioEthanol**
E85

Harvest
BioEthanol

**unleaded**

unleaded

**diesel**

diesel

LOOK HERE

**M**
MORRISONS

BioEthanol
E85

**Unleaded**

**Diesel**

Future electric cars will not need to get electricity from anywhere. People are developing new batteries which use a metal called zinc. When zinc mixes with oxygen from the air, a chemical process makes electricity. When the battery is finished, the zinc can easily be recycled and used again.

## Hydrogen

The cleanest fuel is hydrogen. When hydrogen burns, it just makes water. We already have cars and even helicopters that use hydrogen, but this gas is difficult to produce. Today, most hydrogen is made from fossil fuels, and this produces pollution. About 4 per cent of hydrogen is made from water, but this process uses a lot of electricity and can be dangerous.

In 2010, engineers invented a machine which uses energy from the sun to make hydrogen. Other people are trying to use bacteria to make it. These 'clean' ways of making hydrogen are still very new. If they become cheaper, we may all drive hydrogen cars in the future.

Scientists are developing lots of new fuels, but it will be a long time before most of us can use them. The biggest problem is finding the money to make the necessary changes to things like cars and petrol stations. For example, there are about 140,000 petrol stations in the USA today, but only 2,800 of them sell bioethanol.

Renewable energy is energy which comes from things that go on and on, like the sun or the wind. Some of the fuels in this chapter are renewable, and some are not. Biofuels are renewable because we can grow them again every year. Hydrogen is a renewable fuel when we make it from water but not when we make it from fossil fuels. In the next few chapters, we will look at some other types of renewable energy.

# 7      A bright future

The sunlight that reaches earth in one hour has as much energy as all the power that people use in a year! But how can we get this energy and use it on earth?

'Solar' means 'coming from the sun', so when you use sunlight to make things hot, it is called solar thermal power. Many buildings use materials like glass and plastic to catch sunlight and heat the building. In Africa, people use solar cookers. When light hits the surface of the cooker, it is reflected into the middle. The middle becomes hot enough to heat water or cook food. In countries like Turkey and China, people put solar water heaters on their roofs. These are metal and glass boxes with water pipes in them. The glass catches heat and the metal reflects sunlight onto the water pipes, which carry the hot water down into the houses.

We can use sunlight to make electricity too, with devices called solar cells, which are made of silicon. When sunlight hits the

Using a solar cooker
in Mali, Africa

silicon, particles inside it move, and this makes electricity. One solar cell does not produce much power, so we put the cells together to make big solar panels.

At the moment, the best solar cells can only use about 25 per cent of the sunlight that hits them, and they are an expensive way to produce electricity. But people are inventing better and cheaper solar cells all the time. In the future, we will use it to do more and more things. You can already buy solar lights, solar radios, and small solar panels for things like computers and phones.

We can use solar power to travel too. In July 2010, André Borschberg flew a solar plane called *Solar Impulse* for 26 hours before he stopped. Power for the four engines came from 12,000 solar cells on the wings of the plane. It was able

*Solar Impulse* with its dark solar cells

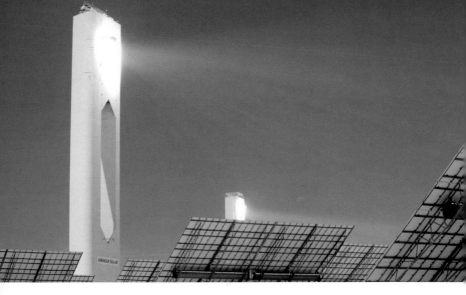

to fly at night because of batteries inside the plane which kept solar energy. There are also solar boats, and every two years, in the World Solar Challenge, solar cars leave Darwin on a 3,000-kilometre journey across Australia. They all try to be the first to arrive in Adelaide, and the fastest cars can reach 100 kilometres per hour.

In sunny countries like Spain, China, and the USA, they are building huge solar power stations. Some use solar panels and others use devices called reflectors to reflect sunlight onto water pipes or tall towers. The Andasol power station in Spain is as big as seventy football fields. It produces enough energy for 200,000 homes!

Imagine standing in the Sahara Desert in fifty years' time. The bright sun hurts your eyes and the heat is fantastic. All around, you can see tall towers and thousands of solar reflectors. It is only a dream at the moment, but many people want to build hundreds of solar power stations in the Sahara Desert, where it is hot and sunny for 365 days a year. Just 0.3 per cent of the Sahara Desert gets enough sunlight to produce electricity for all the people in Europe!

# 8  When the wind blows

On a clear day, you can see them from the land. They look like huge metal flowers growing out of the sea. When you get closer, you realize how big they are. The Thanet Wind Farm is 12 kilometres off the English coast. Each of its 100 wind turbines is 115 metres tall. As wind turns the turbines, generators inside them produce electricity. Together the turbines make enough power for 200,000 homes. There are 250 wind power stations, called wind farms, in Britain, and people are building more every year. By 2020, Britain may get 25 per cent of all its energy from the wind.

The Thanet Wind Farm, England

Bahrain is famous for the strong winds that blow at different times of the year. The World Trade Centre in Bahrain is a skyscraper with three wind turbines. The turbines are 29 metres across, and they produce about 15 per cent of the electricity for the building: that is enough energy for three hundred homes. The shape of the building moves the wind towards the turbines.

It was very difficult to build the Bahrain World Trade Centre because it was the first building of its type in the world. The engineers had to stop the moving turbines shaking the building and destroying it. Now, skyscrapers with huge wind turbines are appearing in cities around the world.

Some future skyscrapers will not use turbines: the building will turn in the wind. As the parts of the building move, generators between the floors will produce power. Because the wind will move faster at the top and slower at the bottom of the building, each floor will turn in a different way. What you can see from your window will change all the time!

Lots of city buildings now have small wind turbines on the roof, and you can even buy one to power your house. Wind turbines used to be ugly, but now there are lots of different shapes and colours and they are great for parks and city centres. Soon tourists will see a wind turbine with changing colours and pictures outside Buckingham Palace in London, for example.

Of course, wind turbines do not work when the wind is not blowing, but people are finding new ways to catch or produce enough wind. When it is not very windy on the ground, look at the sky, and you will often see the clouds moving fast above you. MARS wind turbines fly a few hundred metres above the ground. Their shape makes them turn in the wind, and that produces electricity. Long cables tie them to the ground and carry the electricity. Another idea is to use wind turbines next to busy roads. As cars go past, they move a lot of air, and this drives the turbines.

In Jinshawan, China, a tall tower stands in a field of glass. The sun heats the air under the glass. The hot air moves up the tower and wind turbines inside the tower make electricity. In the future, people are planning to build towers like this in places like Australia and the USA. Some of them

The Greenbird

will be higher than the tallest skyscrapers in the world!

We have used the wind to sail for thousands of years, but the engines of today's big ships use diesel fuel, which comes from oil. However, some new passenger ships have both engines and sails. The sails on these modern ships are moved by computers so they can move and catch the wind, saving the ship a lot of fuel. Skysails are another great idea because you can use them with any ship – even old ones. They fly about 200 metres above the ship and help to pull it through the water.

How about sailing to work in a car that uses the power of the wind? The Greenbird looks like a plane or a boat, but it moves on land. Sometimes it can reach 200 kilometres an hour! The Greenbird uses clever technology to move five times faster than the wind.

An old windmill

The Netherlands is a country of windmills – old buildings with sails that turn in the wind. The country's 1,200 windmills are hundreds of years old, and they are used for lots of things, like moving water and making machines work. Now, wind turbines that make electricity are appearing all over the Netherlands. For the people of the Netherlands, wind power is both the past and the future.

# 9        Water world

I am standing by the Seyhan River, in Turkey, and watching fish swimming in the water. How many rivers like this are there in the world? Think of all that water running down towards the sea. Moving water has much more energy than the wind, and of course, it never stops.

## Hydroelectric power

A hydroelectric power station uses the power of water to produce electricity. The Itaipu Dam, between Brazil and Paraguay, is one of the largest hydroelectric power stations in the world. It took 40,000 workers nine years to build it, and engineers had to move the Paraná River – one of the greatest rivers on earth. The Itaipu Dam is as tall as a skyscraper. Behind a dam, water is kept in a lake called a

The Itaipu Dam

reservoir. Water from the reservoir moves down through huge pipes in the dam. The water turns turbines, which turn generators to produce electricity. The Itaipu Dam produces 90 per cent of Paraguay's electricity as well as enough power for 600 million people in Brazil!

Dams can be good in many ways. In Egypt, there used to be terrible floods on the River Nile after heavy rain, but the Aswan Dam has stopped the floods. Water stays behind the dam in Lake Nasser, and farms can use this water when it is dry. In Turkey, five dams on the Euphrates River have changed the land around them. Farmers can grow much more food, and people from the cities spend their weekends enjoying the beautiful lakes.

The biggest problem with dams is that their reservoirs cover a lot of land. Beautiful places and old buildings often disappear under the water, and people have to leave their homes. After engineers built the Itaipu Dam, people in boats worked hard to save animals as the water flooded the forest. They also moved thousands of trees and plants to higher places.

At Abu Simbel, in Egypt, people saved two 3,000-year-old Egyptian buildings from the waters of Lake Nasser. The huge buildings were cut into pieces, then built again in a higher place. The workers had to cut a lot of stone by hand, and some of the stones were 30 tonnes!

You do not have to build a huge dam to get energy from rivers. A lot of people in Africa and Asia make hydroelectric power themselves. They put pipes and little turbines into rivers in places where the water is moving fast. This produces enough electricity for one or two houses, or sometimes a village. The smallest turbines only cost about 20 dollars: there are more than 100,000 of them in rivers in Vietnam. Of course, we do not have to get power only from rivers; 97 per cent of the world's water is in the sea. Now we are beginning to use it.

## Power from the sea

In 2008, passengers in planes flying to Porto, Portugal, looked down and saw three strange shapes in the water a few kilometres from the beach at Aguçadoura. Each of these big red shapes was 142 metres long. In fact, this was the world's first wave farm, and the strange shapes were wave power generators. In the future, we may see many more farms like

Wave power generators

these near the Portuguese coast, getting power from the restless waves of the Atlantic.

When wave power generators go up and down on the water, liquids inside them move. The liquids turn turbines or push other devices to generate power. Long cables under the sea carry the electricity to the land. The great thing about wave power is that you can get electricity 24 hours a day. The problem is that the waves can be too big. Storms can destroy the machines or break the cables.

Getting power from the waves is still very difficult, but people are developing new devices all the time. In May 2010, a ship pulled a 200-metre wave power generator through Atlantic waters to Orkney, north of Scotland. In a few years, sixty-five more machines will join it, and together they will produce power for about 30,000 homes. Building and using the generators will mean new jobs for hundreds of people in Scotland.

Near New York City, six turbines sit at the bottom of the East River. They are 5 metres across and they look like wind turbines, but they get energy from sea water that moves past them twice a day. In places where rivers join the sea, the water can go up and down 10 metres or more, so it moves very fast and has a lot of energy. To use this energy, we can put turbines under the water or build dams. The world's biggest dam that uses the power of the sea is at La Rance, in France. It is more than 300 metres long and its twenty-four turbines produce enough power for a city of more than 200,000 people.

Near the coast of Kona, Hawaii, the weather is sunny and the sea is very deep. The water is warm at the surface, but 1,000 metres below this it is 10–20 °C cooler. People use this temperature difference to make power. Power stations like the one at Keyhole Point in Hawaii use liquids like ammonia that become a gas at low temperatures. Warm sea water heats the liquids so they turn into gas. The gas moves turbines to generate electricity. Colder water from deeper under the sea turns the gas back into a liquid, so we can use it again.

Many countries already get a lot of their energy from water. For example, Norway gets 99 per cent of its electricity from hydroelectric power. We have used hydroelectric power for years, but in the future, people will get more energy from the sea too. We are still learning how to do this, but there are lots of possibilities. After all, 70 per cent of our planet is covered by water!

Putting a turbine into the East River

# 10   Heat all around us

Imagine you are swimming outside in the beautiful blue waters of a hot pool near Reykjavik, Iceland. Snow is falling all around you, but the water is warm. After a relaxing swim, you catch a bus into the city, then walk home through the streets. It is very cold, and the trees are heavy with snow, but there is no ice on the pavement. Why? Because under your feet, hot water is heating the streets.

Geothermal energy comes from heat under the ground, and people have used it for thousands of years. The Romans used geothermal water to heat bath houses. In New Zealand, Japan and Iceland, people enjoy swimming in geothermal pools.

Iceland has cold winters and short summers, but it is also a land where hot water and steam come up from under the

A geothermal pool and power station in Iceland

ground. In some places, the steam is 250 °C! Iceland's five geothermal power stations use the steam from wells to drive turbines and produce about 25 per cent of the country's electricity. In other places, machines called pumps take hot water from the ground and send it through pipes to houses, and 87 per cent of the buildings in Iceland get their hot water and heating in this way. Hot water under the roads and pavements keeps them clear of snow and safe in the winter.

In places like Iceland, steam comes out of the ground naturally. In other places, pumps send cold water down through pipes to hot dry rocks hundreds of metres below. The rocks heat the water and make steam. A second pipe takes the steam from under the ground.

How safe is geothermal power? The answer is that it is not usually dangerous, but it can be. It is often difficult to drill geothermal wells. The steam can explode from the well, and it can bring dangerous gases with it too. Pumping cold water into the ground is not always safe either. In Basel,

A heat pump for a UK building

Switzerland, a geothermal power station which used cold water was closed after only six days, because in that time, there were 10,000 small earthquakes!

Today we only get geothermal energy from places where very hot rock lies close to the surface. However, the rocks become hotter as you go deeper everywhere on earth, because the centre of the earth is about 5,500 °C! The problem is that in most places you have to go down about 10 kilometres to find enough heat. Some oil wells are already this deep, but it is very expensive to drill them, so people do not do it yet for geothermal wells.

How hot is the ground under your feet? In most places around the world, if you dig down about 3 metres, the ground temperature is 10–16 °C all year. Devices called heat pumps use this heat to turn a special liquid into a gas. The gas moves through pipes next to water. As the gas loses heat energy and becomes a liquid again, it heats the water. In cold countries, hot water from heat pumps is used to heat buildings like schools, houses, and swimming pools.

When some materials become hot, particles inside them move, producing electricity. These are called thermoelectric materials. On a summer day, the streets of a city become very hot: a road can be 70 °C! In the future, some people want to put thermoelectric materials under the roads. This could change the heat into electricity for things like street lights.

# 11 People power

In 1988, a plane called *Daedalus* flew 115 kilometres between the Greek islands of Crete and Santorini. It is a very short flight for today's aeroplanes, but this plane did not have any engines. The power for the plane came from the pilot; he used his legs like someone riding a bicycle to make the plane go forward. In a time before modern machines, people used the power of their bodies to build the Great Wall of China. Today, people power is back.

*Daedalus* in the air

In a small village in Malawi, Africa, children shout excitedly as they play on a merry-go-round. It is the favourite meeting place for all the village children. As they turn around and around, a pump uses their movement to bring water up from a well under the ground.

Children enjoying
a play pump

In Africa, getting clean water is a problem for many people. They may be many kilometres away from rivers, and river water is not always clean. Getting clean water from under the ground can be difficult, because pumps with engines are expensive to use and they often break. These merry-go-round 'play pumps' mean that villages and schools can have clean and safe drinking water – and the children can have fun too!

When you run, you have a lot of energy which comes from the movement of your body. When you suddenly stop, your body loses this energy. We already have watches and small medical devices which can use energy that we make when we move. In the future, people like police officers and soldiers may wear devices on their legs to 'catch' this lost energy and keep it in batteries. They could use the power for computers, radios or other devices.

In December 2008, most people walking across Hachiko Square, Tokyo, probably did not notice four yellow squares on the pavement as they hurried to work. The squares were made of special materials that make electricity when they change shape. When people stood on the squares, the shape of the materials changed and they produced electricity. The squares were only there for twenty days, but in that time they produced enough power to make a TV work for 1,400 hours! Imagine putting these squares under all the roads and pavements in Tokyo. One day, we may turn our streets into power stations!

Moving people can produce a lot of energy, but what about people who cannot move about – sick people or people sitting on trains? Even when we are resting, our bodies produce enough energy to power two laptops! Most of this energy is heat. Now people are developing medical devices which get their power by changing body heat into electricity. Soon doctors will use them to do things like getting information about their patients' blood, for example. This will be useful in places like Africa, where many villages do not have electricity.

We can use body heat in other ways too. Every day, 250,000 people use Stockholm's Central Station. They eat and drink, carry heavy bags, and run to catch trains – and their bodies produce a lot of heat when they do these things. Inside the station, heat pumps take heat from the air and use it to heat water for a nearby building. It is a great way to get free energy – all you need is a lot of people!

# 12            Nanopower

In 2005, scientists in the USA built a tiny car called the Nanocar. How small is it? It is difficult to imagine something so small, but a hair on your head is 20,000 times wider and millions of times longer! The wheels of the Nanocar were made from balls of carbon atoms. Scientists used heat to move the wheels and 'drive' the car.

Nanotechnology is building things from atoms or from molecules (which are atoms joined together, like $H_2O$). In the future, we will use it to make tiny machines called nanobots. Millions of these machines will do things like clean waste and build or fix things. Doctors may use them to help sick people: they could travel through our bodies and fix damage inside us.

Tiny generators called nanogenerators will produce power. They will get energy from light or from movements around them: for example, blood moving around your body or sound moving through the air. In the future, we may print nanogenerators onto materials for making clothes. They may power the phones in our pockets, heat our clothes in cold weather, or change the colours of a favourite dress.

Scientists have already invented the first nanogenerators: tiny solar cells. Most solar cells only use the light that people can see. Nano solar cells can also use a different kind of light called infrared light; hot things produce this light all the time, even in the dark. We already use nano solar cells for making thin solar panels, but they are still very new. One day we may add them to liquids which we can put onto the outside of houses and cars. They will use the electricity that the nano solar cells produce.

While some people are making solar cells which are too small for us to see, other people want to put huge solar panels into space. We will read about them in the next chapter.

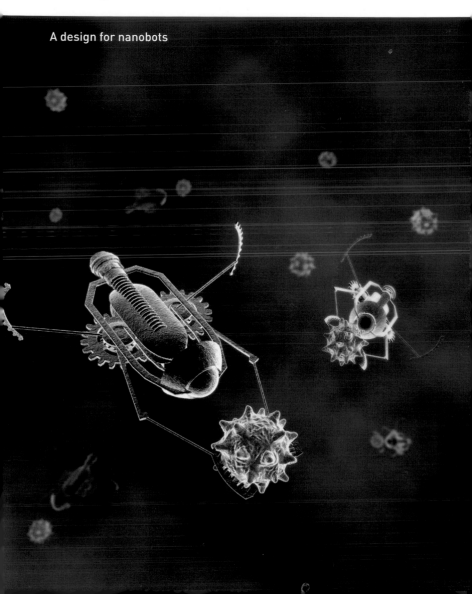

A design for nanobots

# 13 Energy in space

Imagine that the date is 2095. You live in a world without oil. Far above you, huge shapes are moving across the sky, but you cannot see them from the ground. Like many places in the world, your city gets a lot of its energy from power stations in space. It does not sound possible, but it may happen one day.

At work on a solar panel in space

We have used solar panels in space for years. For example, they power the satellites that move around earth and send us information about the weather. Now some people want to send solar energy back to earth. How will they do this?

They plan to make huge solar panels from very light materials and pack them into small containers. When they are in space, they will open and join together to make solar power stations. Space solar power stations will take energy from the sun and send it back to earth, where we will change it into electricity.

Gases and clouds stop a lot of solar energy before it

reaches the earth, but this is not a problem for solar panels in space. If they move around the planet, they can stay where the sun is and work 24 hours a day. A company in California plans to put the first panels into space in 2016. They will produce enough electricity for 250,000 homes.

When we send spacecraft to other planets, we often have to find unusual ways to power them. Some have engines which produce special atoms called ions to push them forward. Others use nuclear fuel. In 2010, a Japanese spacecraft called IKAROS began 'sailing' towards Venus. When particles from the sun hit its 'solar sails', they pushed the spacecraft forward, like the wind pushes the sails of a ship. Because of the success of IKAROS, other spacecraft may use solar sails in the future.

For more than ten years, scientists from around the world have lived and worked in space on the International Space Station. In the future, countries like China, Russia, and Japan plan to build places to live and work on the moon. People will need to produce energy there because they cannot take enough fuel from earth. They can use solar power in the day, but nights on the moon are nearly two weeks long! How will they get energy to use in the dark?

At NASA, scientists have invented special batteries to use on the moon. In the day, the batteries use solar energy to get hydrogen and oxygen from water. In the night, the battery mixes these gases so that they burn. This produces energy and leaves water, which the batteries can use again.

Since the Apollo 17 spacecraft left the moon in December 1972, nobody has visited the moon. One day, people will go back – and this time, with the help of new devices, they will probably stay.

# 14 Going local

We do not realize it, but we use space technology all the time. When we watch TV, the pictures may come from a satellite hundreds of kilometres above the earth. It is the same when we make phone calls or listen to the radio. Far above our heads, machines are working for us, but we will never see most of them.

A few hundred years ago, people mainly used the things they could see around them. Local people grew the food and made the products that people needed. For fuel, most people got wood from a nearby forest. Today, the things that we buy and the fuel that we use often get to us from the other side of the world. This wastes energy – and what happens when these things do not come?

## Local power

In October 1973, governments argued, and countries in the Middle East stopped selling oil to Europe and the USA. Very soon, life started to change. In the USA, drivers waited for hours to buy petrol.

In Europe, people did not have enough fuel to heat their houses. The problems only lasted for five months, but companies closed and thousands of people lost their jobs. What did we learn from this? The short answer is probably 'not very much'.

Most countries still get most of their energy from fossil fuels. This often comes in pipes or ships from thousands of kilometres away. In 2009, Russia and the Ukraine argued

**'No gas' (petrol) in the USA, 1973**

about the price of natural gas, so people stopped pumping it. In some places in Europe, people had no heating in the cold winter, and the temperature was down to -10 °C!

Slowly, different countries are starting to produce their own energy in different ways, so they need less fuel from other places. Brazil already produces a lot of biofuels, and Norway and Iceland get most of their power from hydroelectric and geothermal energy. In the future, sunny Spain may get a lot of power from solar energy, and stormy island countries like Britain will use wind turbines and wave power generators.

Today we produce electricity in huge power stations and send it through cables to places far away. This wastes a lot of energy. Big power stations lose heat, and more energy is lost when the electricity travels through the power cables. Surprisingly, about 66 per cent of the energy from burning fossil fuels in power stations never reaches our homes.

Sometimes a group of power stations stop working or big power cables burn. When this happens, the lights of big cities may go out. Trains stop running and people sleep in their offices because they cannot go home. So what can we do about these problems? Is there a better way to make and send electricity? We may not find an answer to all these problems, but producing more power locally will help.

In the future, we may produce some of our electricity nearer to our homes. As new energy technologies become cheaper, we will use them in more and more places. People may get their electricity from a wind turbine in their street or a wave power generator at a local beach, and not from a big power station on the other side of the country.

Our houses will make more electricity too. Many will have solar panels or wind turbines on their roofs. We may also put tiny turbines in kitchen and bathroom water pipes. Heat pumps in our walls may make hot water for the house, and perhaps the floors under our feet will make electricity when we walk on them. Each of these devices will only make a little power, but when we put them together, they will make a lot.

## Making things locally

Why is it better to buy books and music that go straight to a computer instead of onto paper or plastic? The answer is that it saves a lot of energy. It saves fuel and materials for making the products and their containers as well. In the future, some companies will save energy by making things in a different way.

Recently we have invented printers which print real objects from computer designs. Now they can even print objects with moving parts. Instead of making things like sports shoes and phones in factories and sending them to other countries, companies may sell designs for them. People will buy the designs and print the products at local 'object printing shops', or even at home.

An object printer

Object printing will be great for small products, but what about making something big, like a car? In 2010, visitors to a car show in Los Angeles were shown a very different way to make cars. There were extraordinary designs for cars made from extra-light and extra-strong materials. The most

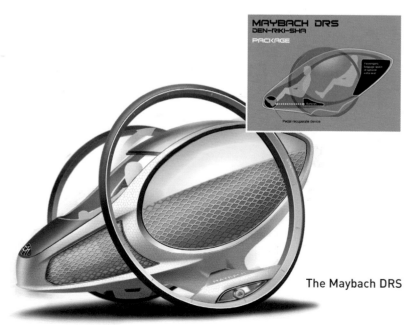

The Maybach DRS

surprising idea was the Maybach DRS: a future car which will make itself from living materials. It will grow like a plant!

If we can invent living materials, they will get energy from the sun in order to grow. Scientists will develop special ways to 'tell' these materials to grow into the parts of a car, or something else. Slowly the parts will appear, like fruit growing on a tree. Instead of sending heavy metal car parts in ships, companies will buy and sell the information that they need to grow the different parts.

When we throw away living materials, bacteria in the ground will break them down into pieces – they will not produce any rubbish or pollution. In the future, people may want to grow many different products, like chairs and houses. Living materials are still just an idea: we have not made them yet. But one day, our factories may turn into farms!

Imagine a world where products change when we want them to! It sounds like something from a Hollywood film, but a lot of people are studying this idea in universities around the world. They think that one day, millions of nanobots will join together and make themselves into objects at home. Each nanobot will be too small for our eyes to see. The objects will change when we want them to: computers will tell the nanobots to move and become a different shape. If this happens one day, we will not need to throw anything away. If there is an object that we do not need, we will make it into something else.

Imagine a desk which becomes a chair when you do not need to work, or a coat which gets longer when it rains. It sounds impossible, but some scientists believe that it will happen in fifty to a hundred years!

# 15 Where next?

Today, about 7 billion people live on earth. In 2050, the population will be more than 9 billion. This means millions more cars, TVs, fridges, and computers – and many other machines that we have not invented yet. Where will we get the energy for all these things?

In this book we have read about lots of ways to produce and save energy. Some ideas are very old. Others are quite new, and some of them look far into the future – to a time when objects can make themselves, and people are living in space.

Which energy technologies will we use most in the future? We do not know the answer to this question, but we do know that we cannot use fossil fuels. Ten years ago, people argued about the idea that pollution from fossil fuels was changing the world's climate. Today, scientists agree that it is – and they think it is happening very fast!

We are starting to build a future without fossil fuels. Turbines are appearing in rivers and on hills around the world, and people are starting to use new fuels. Every year, scientists are inventing new ways to make electricity from the things around us. We know a lot of ways to save energy, and some of us can even try to produce it ourselves. The big question is 'Can we change in time – before we destroy a lot of the natural world?' And the answer? We will have to wait and see.

Crowds in Shanghai, China

# GLOSSARY

**amount**  how much there is of something

**atom**  one of the very small things that everything is made of

**bacteria**  very small things that live in air, water, earth, plants, and animals

**battery**  something that makes electricity for a clock, radio etc.

**cable**  a strong thick metal rope

**chemical**  something solid or liquid used in chemistry

**coal**  a hard black substance that comes from under the ground and gives out heat when you burn it

**coconut**  a large brown hard fruit that grows on trees in hot countries

**container**  a thing that you can put other things in, e.g. a box

**cool**  to make something less hot

**dam**  a wall that is built across a river to stop the water

**design**  a plan that shows how to make something

**develop**  to think of a new idea or product and make it successful

**device**  a tool or piece of equipment that you use for doing a special job

**drill**  a tool that you use for making holes

**earth**  the world

**earthquake**  a sudden strong shaking of the ground

**electricity**  power that can make heat and light

**flood** *(n & v)*  when there is a flood, a lot of water covers the land

**fuel**  anything that you burn to make heat or power;
   **fossil fuel** fuel like coal or oil that comes from animals or plants millions of years old

**gas**  something like air that you burn to cook or make heat

**generator**  a machine for producing electricity

**geothermal**  connected with natural heat that comes from under the ground

**government**  a group of people who control a country

**heat pump** a machine that can move heat from one place to another

**huge** very big

**hydroelectric** using the power of water to make electricity

**invent** to make something that did not exist before

**light bulb** the glass part of an electric lamp that gives light

**liquid** water, oil, and milk are all liquids

**local** of a place near you or where you live

**material** what you use for making or doing something

**movement** when something moves from one place to another

**nano-** very, very small; **nanobot** a very small machine that works by itself; **nanogenerator** a very small machine that produces power; **nanotechnology** a kind of technology that works with very small things

**normal** not different or special

**object** a thing that you can see and touch

**oil** a thick liquid from under the ground that we use for energy

**particle** a part of an atom

**pavement** the part at the side of a road where people can walk

**petrol** the liquid that you put in a car to make it go

**planet** the Earth, Mars and Venus are all planets

**plastic** a light strong material used to make many different things

**pollute** to make the air, rivers etc. dirty and dangerous; *(n)* **pollution**

**pool** a place where people can swim

**population** the number of people who live in a place

**power** *(n & v)* strength; the supply of electricity; to supply energy or electricity to something

**power station** a building where electricity is produced

**print** *(in this book)* to put a design into a machine and produce an object; *(n)* **printer**

**process** a number of actions, one after the other, for doing or making something

**produce** to make something as part of a process

**pump** *(n & v)* a machine that moves water, gas, or air

**radiation** powerful and dangerous rays that come from something radioactive

**recycle** to do something to materials like paper and glass so that they can be used again

**reflect** to throw back light from a surface; *(n)* **reflector**

**satellite** a device in space that moves around the earth and sends back information

**solar** coming from the sun; **solar cell** a device that changes light from the sun into electricity; **solar panel** a lot of solar cells working together

**space** the place beyond earth where the moon and stars are; **spacecraft** a vehicle that travels in space

**standby / on standby** ready to start working immediately

**steam** the gas that water becomes when it gets very hot

**surface** the outside or top layer of something

**tanker** a ship or lorry that carries oil or gas

**technology** using science to build and make new things

**temperature** how hot or cold something is

**tiny** very small

**tonne** one thousand kilograms

**tower** a tall narrow building

**turbine** a machine that gets its power from a wheel that is turned by water or air

**waste** to use more of something than is necessary or useful; *(n)* things that are not wanted or needed

**wave** a raised line of water that moves across the sea

**well** a deep hole in the ground where people get water or oil

# Future Energy

## ACTIVITIES

ACTIVITIES

## *Before Reading*

1 **Match the chemicals with the definitions.**

*Aluminium (Al) / Carbon dioxide ($CO_2$) / Carbon monoxide (CO) / Hydrogen (H) / Methane ($CH_4$) / Oxygen (O) / Silicon (Si) / Uranium (U) / Zinc (Zn)*

1 This light metal is used to make things like planes.
2 This gas can kill you.
3 This is a fuel in a nuclear power station.
4 This is the lightest gas and the smallest atom.
5 This is a gas which people use to cook.
6 This gas is made when things burn.
7 This gas is something we need to live.
8 This metal is found in your body and used in batteries.
9 This is used to make an important computer part.

2 **How much do you know about energy? Three of these sentences are true. Which are they?**

1 The first factories used machines called steam engines.
2 Most of the world's energy comes from nuclear power.
3 You can make fuel for cars from plants.
4 People did not use wind power until the 1970s.
5 Iceland gets most of its energy from burning coal.
6 Your body produces enough energy for a 100W light.

ACTIVITIES

## *While Reading*

**Read Chapters 1 and 2 and circle the correct words.**

1  A plane's engines burn a lot of *fuel* / *energy*.
2  At this moment, billions of machines are *produced* / *turned on*.
3  *Generators* / *Turbines* change movement into electricity.
4  Karl Benz used a *steam* / *petrol* engine in the first car.
5  Fossil fuels come from *rocks* / *living things*.
6  The Romans used coal to make *metals* / *chemicals*.
7  We can use tankers and *wells* / *pipes* to move oil and natural gas around the world.

**Read Chapters 3 and 4 and find numbers to complete the sentences.**

1  Air pollution kills about _____ million people a year.
2  There is _____ per cent more $CO_2$ in the air today than when the steam engine was invented.
3  _____ types of animals and plants disappear every day.
4  The world uses about _____ billion litres of oil daily.
5  Each family in Europe wastes more than _____ dollars of food every year.
6  People in Greece throw away _____ million plastic bottles every day.

**Read Chapter 5. Rewrite these untrue sentences to make them true.**

1  Nuclear fission happens when atoms join together.
2  Control rods help to heat the fuel inside the reactor.
3  The water inside a nuclear reactor changes into steam and turns steam turbines.
4  Most nuclear waste is recycled and used again.
5  An accident happened at Chernobyl after a wave destroyed machines in a nuclear power station.

**Read Chapter 6 and complete the sentences with these words.**

*biofuel / biogas / hydrogen / zinc*

1  _____ is very easy to recycle and use again.
2  It takes a lot of land to produce _____.
3  Bacteria help to make _____.
4  _____ is often made from fossil fuels.
5  People in India use _____ fuel for cooking and lights.
6  _____ makes electricity when it mixes with oxygen.

**Read Chapters 7 and 8. Choose the best question-words for these questions, and then answer them.**

*How / How many / How much / What / Where / Which / Who / Why*

1  . . . do solar cookers work?
2  . . . do we call lots of solar cells together?
3  . . . used solar energy to fly a plane?
4  . . . may we see lots of solar power stations one day?
5  . . . electricity does the Thanet Wind Farm produce?
6  . . . turbines are there at the Bahrain World Trade Centre?
7  . . . do people want to put wind turbines near roads?
8  . . . European country has used wind power for a long time?

**Read Chapter 9, then match these halves of sentences.**

1 The lake behind a dam . . .
2 Water moves through huge pipes in a dam . . .
3 Dams can help to stop . . .
4 People sometimes put small turbines into rivers . . .
5 We can get power from the waves with . . .
6 Wave power generators can . . .
7 Special liquids . . .
8 The water can go up and down 10 metres or more . . .

a) flooding and keep water for when it is dry.
b) be destroyed by bad weather.
c) in places where rivers meet the sea.
d) and turns turbines to make electricity.
e) generators which go up and down on the water.
f) are used to get heat energy from the sea.
g) is called a reservoir.
h) to make power for a few homes.

**Read Chapters 10 and 11 and write the names of the places.**

1 The pavements of this city are heated with hot water.
2 Swimming in geothermal pools is popular in these three countries.
3 There were a lot of earthquakes in this city after people drilled a geothermal well.
4 A man flew an unusual plane to this island.
5 In this country, children pump water when they play.
6 When people walked on yellow squares in this city, they made electricity.
7 A building in this city uses heat from people's bodies.

**Read Chapters 12 and 13, then fill in the gaps with these words.**

*atoms, infrared, nanogenerators, satellites, solar sails, space station, sun*

1  The Nanocar was made from carbon _____ .
2  _____ are very small machines which can produce power.
3  _____ is a kind of light that we cannot see.
4  _____ move around the earth in space.
5  In space, solar panels can work all the time because they move to where the _____ is.
6  _____ use particles from the sun to push a spacecraft.
7  In space people live and work in a _____.

**Read Chapters 14 and 15. Are the statements true (T) or false (F), or not mentioned (N)?**

1  Normal life becomes very difficult when there is no oil.
2  Russia sells most of its natural gas to Europe.
3  More than half the electricity made in power stations is lost before it arrives at people's houses.
4  In the future, our houses may produce power.
5  Computers are inventing designs and printing objects from them.
6  In Los Angeles people have started growing cars.
7  Soon, India's population will be bigger than China's.
8  We are not sure that pollution is changing the weather.
9  Scientists are finding new ways to use fossil fuels to make electricity.

## *After Reading*

**1 Complete these two news stories with the words below.**

*electricity, fission, fuel, fusion, homes, million, panels, problems, reactor, reflect, safer, steam, sunlight, temperatures, turbines, waste*

---

### SOLAR POWER LIGHTS HOUSES IN ANDALUCIA

A huge new solar power station has opened near Seville, in Spain. The 330-million-euro power station's 2,650 glass _____ catch _____ and _____ it onto a tower. Inside the tower, heat is used to turn water into _____, which turns _____ and produces enough _____ for 25,000 _____ in Andalucia.

---

### MIT FINDS A WAY TO CATCH THE SUN

Nuclear _____ happens naturally on the sun at _____ of about ten _____ degrees – too hot for any container on earth. Now scientists at MIT have discovered a way to keep the super-hot _____ used in nuclear fusion from touching the sides of the _____ and destroying it. Nuclear fusion is much _____ than nuclear _____, and it produces less radioactive _____. One day it could be the answer to all of our energy _____.

2  **Complete the crossword. Then use the letters in the shaded boxes to make two mystery words.**

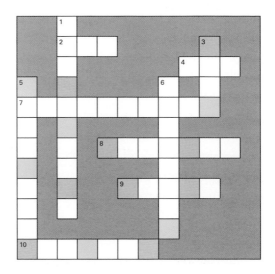

**Across**

   2  A thick, black liquid that is made into petrol.
   4  A wall across a river which stops the water.
   7  When the ground shakes suddenly.
   8  This makes electricity for a clock or radio.
   9  Near to where you live.
   10  To change materials like paper and glass and use them again.

**Down**

   1  The people who control a country.
   3  A moving line of water.
   5  A machine which makes electricity.
   6  Part of an atom.

**The mystery words are:** _ _ _ _ _ _ _ _ _   _ _ _ _ _ _

3 Do you think these things will happen in the next 100 years? Write 1–5. (1 = definitely not, 2 = probably not, 3 = perhaps, 4 = probably, 5 = definitely)

1 Pollution will kill 30 per cent of all the world's animals.
2 New York and Shanghai will disappear under the sea.
3 Most of the electricity that you use will be made in your street or your house.
4 The furniture in your bedroom will change shape and colour when you want.
5 We will recycle 95 per cent of all our rubbish.
6 Factories will 'grow' many products like plants.
7 People will begin to live on the moon.

4 Do you agree or disagree with these sentences? Why?

1 How does your country get most of its energy?
2 How can governments help people to use less energy?
3 Which is more important for our future: finding ways to save energy or new ways to produce it?
4 Which forms of energy will people use the most fifty years from now?
5 How will new energy technologies change our lives?

5 Think about how we can save energy and make a poster or prepare a class quiz. Think about what you do at home, the things that you buy and throw away, and how you travel. These websites can help you:

www.recyclenow.com
www.energyquest.ca.gov
www.eia.gov/kids

# ABOUT THE AUTHOR

Alex Raynham has taught English in Italy, the UK, and Turkey. In the last ten years, he has worked on many books for Oxford University Press, and has written graded readers for Oxford Read and Discover and Dominoes, as well as Factfiles.

He now lives with his wife Funda in the sunny Turkish city of Adana. At the weekend, they often go walking around the lake near their house. It's the reservoir of the Seyhan dam, which produces most of the city's electricity.

Like most houses in Adana, their apartment uses solar water heating for about nine months of the year. Moveable glass panels also close the balconies in winter so that two-thirds of the house is behind glass. The sun warms the air behind the glass, and this keeps the apartment very warm in cold weather. The glass panels are opened in summer to let cool air into the house. At night, all of the machines in the house are switched off, except for the fridge and the telephone.

Alex and Funda have a petrol car, which isn't good for the environment. There are boxes of rubbish in the boot of the car. Why? Because they collect their rubbish and take it to the nearest recycling bins, which unfortunately are 10 kilometres away!

# OXFORD BOOKWORMS LIBRARY

*Classics • Crime & Mystery • Factfiles • Fantasy & Horror*
*Human Interest • Playscripts • Thriller & Adventure*
*True Stories • World Stories*

The OXFORD BOOKWORMS LIBRARY provides enjoyable reading in English, with a wide range of classic and modern fiction, non-fiction, and plays. It includes original and adapted texts in seven carefully graded language stages, which take learners from beginner to advanced level. An overview is given on the next pages.

All Stage 1 titles are available as audio recordings, as well as over eighty other titles from Starter to Stage 6. All Starters and many titles at Stages 1 to 4 are specially recommended for younger learners. Every Bookworm is illustrated, and Starters and Factfiles have full-colour illustrations.

The OXFORD BOOKWORMS LIBRARY also offers extensive support. Each book contains an introduction to the story, notes about the author, a glossary, and activities. Additional resources include tests and worksheets, and answers for these and for the activities in the books. There is advice on running a class library, using audio recordings, and the many ways of using Oxford Bookworms in reading programmes. Resource materials are available on the website <www.oup.com/elt/bookworms>.

The *Oxford Bookworms Collection* is a series for advanced learners. It consists of volumes of short stories by well-known authors, both classic and modern. Texts are not abridged or adapted in any way, but carefully selected to be accessible to the advanced student.

---

You can find details and a full list of titles in the *Oxford Bookworms Library Catalogue* and *Oxford English Language Teaching Catalogues*, and on the website <www.oup.com/elt/bookworms>.

## THE OXFORD BOOKWORMS LIBRARY
## GRADING AND SAMPLE EXTRACTS

### STARTER • 250 HEADWORDS

present simple – present continuous – imperative –
*can/cannot, must* – *going to* (future) – simple gerunds ...

Her phone is ringing – but where is it?

Sally gets out of bed and looks in her bag. No phone.
She looks under the bed. No phone. Then she looks behind
the door. There is her phone. Sally picks up her phone and
answers it. ***Sally's Phone***

### STAGE 1 • 400 HEADWORDS

... past simple – coordination with *and, but, or* –
subordination with *before, after, when, because, so* ...

I knew him in Persia. He was a famous builder and I
worked with him there. For a time I was his friend, but
not for long. When he came to Paris, I came after him –
I wanted to watch him. He was a very clever, very
dangerous man. ***The Phantom of the Opera***

### STAGE 2 • 700 HEADWORDS

... present perfect – *will* (future) – *(don't) have to, must not, could* –
comparison of adjectives – simple *if* clauses – past continuous –
tag questions – *ask/tell* + infinitive ...

While I was writing these words in my diary, I decided
what to do. I must try to escape. I shall try to get down the
wall outside. The window is high above the ground, but
I have to try. I shall take some of the gold with me – if I
escape, perhaps it will be helpful later. ***Dracula***

### STAGE 3 • 1000 HEADWORDS

*... should, may* – present perfect continuous – *used to* – past perfect –
causative – relative clauses – indirect statements ...

Of course, it was most important that no one should see
Colin, Mary, or Dickon entering the secret garden. So Colin
gave orders to the gardeners that they must all keep away
from that part of the garden in future. *The Secret Garden*

### STAGE 4 • 1400 HEADWORDS

... past perfect continuous – passive (simple forms) –
*would* conditional clauses – indirect questions –
relatives with *where/when* – gerunds after prepositions/phrases ...

I was glad. Now Hyde could not show his face to the world
again. If he did, every honest man in London would be
proud to report him to the police. *Dr Jekyll and Mr Hyde*

### STAGE 5 • 1800 HEADWORDS

... future continuous – future perfect –
passive (modals, continuous forms) –
*would have* conditional clauses – modals + perfect infinitive ...

If he had spoken Estella's name, I would have hit him. I was so
angry with him, and so depressed about my future, that I could
not eat the breakfast. Instead I went straight to the old house.
*Great Expectations*

### STAGE 6 • 2500 HEADWORDS

... passive (infinitives, gerunds) – advanced modal meanings –
clauses of concession, condition

When I stepped up to the piano, I was confident. It was as if I
knew that the prodigy side of me really did exist. And when I
started to play, I was so caught up in how lovely I looked that
I didn't worry how I would sound. *The Joy Luck Club*

BOOKWORMS · FACTFILES · STAGE 3

# Formula One

ALEX RAYNHAM

It's an exciting life – full of fast cars, money, and travel. The names of Formula One champions are known all over the world. And everywhere young drivers dream of success one day in Monaco, Melbourne, Monza . . .

But it is a difficult life too. Drivers need strong bodies – and minds. They need to think quickly, drive hard, and sometimes look death in the face. This is the dangerous, exciting world of Formula One – where the world's best drivers have only seconds to win or lose a race.

BOOKWORMS · FACTFILES · STAGE 3

# Recycling

SUE STEWART

What will we do when there is nowhere to put our rubbish? Every day, all over the world, people drop cans, boxes, paper, and bottles into bins and never think about them again. And the rubbish mountains get bigger and bigger.

But there is another way – a way that makes old paper into houses, broken bottles into jewellery, and old cans into bridges. Anyone can recycle – it's easy, it saves money, and it's a way to say, 'I care about the Earth.' Saving the world starts with you – here – now.